I0062906

THE FUTURE OF

CYBERSECURITY

What's Coming Next and How to Prepare

Author

Fardis Enayat

Author's Note

This book was written to help make sense of a changing reality.

After years of working in IT and cybersecurity, I noticed that

many of the most damaging incidents were no longer driven by

complex technical exploits. Instead, they were caused by

believable **messages, rushed decisions, and misplaced trust.**

Systems weren't *always broken; people were persuaded.*

My first book, "Hacker Vs Security Expert," explored the

technical side of cybersecurity and how attackers and defenders

operate within digital systems. This book takes a different

approach. It focuses on the human layer: how trust is formed,

how it is exploited, and why modern cyber threats feel so

ordinary until it is too late.

The ideas in this book are shaped by real-world experience,

observation, and ongoing conversations with professionals who

deal with digital risk every day. While technology continues to

evolve, the core challenge remains the same making sound

decisions under pressure.

If this book helps you slow down at a critical moment, question something that feels normal, or build systems that reduce the impact of mistakes, then it has served its purpose.

Fardis Enayat

Dedication

Thirteen years ago, when I came to the United States, I could not speak or understand English well. Communicating my thoughts was difficult, and believing in myself was even harder. Through every challenge, my wife stood by me. She helped me believe in myself when I doubted, supported me when things were uncertain, and stayed strong with me through our darkest moments. This book would not exist without her patience, encouragement, and constant belief that I could grow beyond where I started.

Why This Book, Why Now

Cybersecurity didn't suddenly become important.

It quietly became personal.

For a long time, cyber threats felt distant to most people.

Something that happened to governments, banks, or giant tech companies. Something technical. Something handled by "IT."

That world is gone.

Today, cybersecurity lives inside everyday moments:

- An urgent message from a boss
- A call that sounds like someone you trust
- A login prompt you didn't expect
- A payment request that feels routine
- A system that assumes you're real because your credentials worked

The danger isn't that attacks are louder than before.

The danger is that they feel normal.

Why I Wrote This Book

In my first book, *Hacker vs Security Expert*, I focused on the technical battle, how attackers think, how systems are compromised, and how defenses try to keep up. It was written for readers who wanted to understand the mechanics of modern cyber warfare.

But after years in IT and cybersecurity, I began to notice a different pattern.

Some of the most damaging incidents weren't caused by brilliant technical exploits.

They were caused by:

- Believable messages
- Rushed decisions
- Borrowed trust
- Perfectly reasonable actions taken at the wrong moment

No zero-day exploits.

No advanced malware.

No dramatic "hacking" scenes. Just people doing their jobs.

That's when it became clear: the center of cybersecurity has shifted.

This book exists because **the most dangerous attacks today don't break systems , they exploit trust**.

Why Now Matters

We're at a turning point.

Three forces are colliding at the same time:

1) Cybercrime has professionalized

Attacks are no longer random. They are organized, scalable, and designed like businesses. That means volume, efficiency, and repeatability.

2) AI has raised the realism bar

Messages are clearer. Voices are convincing. Fake requests feel calm and professional. The old warning signs are fading.

3) Our lives are fully connected

Cloud services, remote work, vendors, shared links, and constant authentication mean there are more doors than ever, and many of them look legitimate.

This combination changes the rules.

Awareness alone is no longer enough.

Experience alone is no longer protection.

Being "smart" is no longer immunity.

That's why this book matters *now*, not later.

What This Book Is (and Is Not)

This book is **not** a technical manual.

It won't drown you in tools, frameworks, or jargon.

This book is **not** about turning you into a cybersecurity professional.

Instead, it's about:

- Understanding how modern attacks actually succeed
- Recognizing the patterns that repeat across incidents
- Learning how to slow down the moment's attackers rely on
- Building habits and systems that reduce damage when mistakes happen

This is a book about **survivability**, not perfection.

Who This Book Is For

This book is for:

- Professionals who make decisions under pressure
- Leaders responsible for people, money, or systems
- Individuals who want to protect their families and identities

- Anyone who feels the rules have changed but can't quite explain how

You don't need a technical background.

You just need to live in the modern world.

The Core Idea

If there's one idea that runs through this entire book, it's this:

The future of cybersecurity is not about stronger walls.

It's about protecting trust when trust can be faked.

Everything you'll read, from AI scams to identity attacks, from ransomware to personal security, flows from that reality.

How to Read This Book

You don't need to read this book like a textbook.

Read it like a playbook.

Pay attention to:

- The patterns
- The moments where speed is used as a weapon
- The points where verification would have changed the outcome

You'll see the same ideas come back again and again on purpose.

Because repetition is how habits form.

Where We Go Next

We start with the foundation.

Before we talk about AI, ransomware, identity, or zero trust, we need to understand the shift that makes all of it possible: **Trust is breaking.**

And once you see that clearly, the rest of the future starts to make sense.

Chapter 1

A New Kind of Risk

(Deepfakes, voice scams, fake "boss"

requests)

Trust is breaking.

Not because people are getting "dumber."

Not because technology is evil.

But because the tools for faking reality are getting cheaper,

faster, and more believable.

And the more believable a lie becomes, the more it works on

everyone.

A Tuesday Afternoon That Doesn't Feel Like a Cyberattack

It starts like nothing.

You're in the middle of work. You're answering messages.

You're thinking about dinner, and your phone rings.

It's your boss.

Same caller ID. Same name. Same voice.

He sounds busy, a little stressed, that tone you've heard before

when something urgent is happening. He says there's a

payment that needs to go out today. He says it's sensitive. He says he'll explain later.

And then he adds the sentence that makes most people move fast:

"Please don't slow this down."

No threats. No yelling. No hacking movie stuff.

Just pressure.

Just urgency.

Just a request from someone you trust.

So, you do it.

And later, sometimes, minutes later, you find out your boss never called.

That voice was fake.

The urgency was designed.

And the money is gone.

This is the new kind of risk.

Because the most dangerous cyberattacks don't look like cyberattacks anymore.

They look like normal life.

What Changed

A lot of people still think cybersecurity is mainly about:

- Antivirus

- Firewalls

- Complicated passwords

- "Don't click the link."

That stuff still matters.

But the center of gravity has moved.

The biggest change isn't that criminals learned new code.

The biggest change is that criminals learned something about people:

If you can't break the system, break the people.

A story like:

- "I'm your CEO, I need this done right now."

- "I'm your bank, your account is locked."

- "I'm your coworker. I lost access. Can you help?"

- "I'm your child, I'm in trouble."

The attack isn't the malware.

The attack is the moment.

It's **timing, emotion, and trust**.

And AI makes it easier to manufacture that moment.

Deepfakes and Voice Scams Don't Need to Be Perfect

Here's what surprises people:

A fake voice doesn't need to be flawless.

It only needs to be believable enough in a rushed moment.

Humans don't verify reality the way a computer does. We verify reality by pattern.

- "That sounds like him."

- "This is how she talks."

- "He would ask me, not someone else."

- "This is urgent. This is important. This makes sense."

And most scams are designed to hit you when you're:

- Tired

- Multitasking

- Under pressure

- Afraid of making a mistake

- Trying to be helpful

- Trying to be fast

That's why smart people fall for it.

Not because they're careless.

Because they're human.

The New Attack Surface Is Your Identity

In the older world, criminals had to fight machines.

They had to break into networks, crack passwords, and exploit systems.

Now they often go after something softer, but more powerful: **identity.**

Because identity is the master key to modern life.

If someone can convincingly pretend to be:

- You
- Your boss
- Your vendor
- Your bank
- Your IT department
- Your family member

They can open doors without breaking anything.

And when nothing "breaks," people don't notice until it's too late.

That's why a lot of modern incidents don't start with a technical "hack."

They start with a message that feels routine.

They start with a conversation.

They start with trust.

The Most Dangerous Scams Are the Ones That Feel Polite

A professional scam doesn't feel like a scam.

It feels like someone is doing their job.

It's written cleanly. It's short. It uses the right words.

It doesn't scream "CLICK THIS NOW."

It says:

- "Can you do me a favor?"
- "I'm in a meeting, handle this."
- "I need a quick confirmation."
- "I'm locked out. Can you approve this?"
- "We're on a deadline, don't delay."

This is why the future of cybersecurity is not just technical.

It's behavioral.

It's cultural.

It's about building systems where **one person's rushed decision cannot sink the whole ship.**

Why does this get worse from here

There are three reasons for this kind of attack will spread.

1) The cost of "believable" is dropping

Years ago, impersonation took time. It took effort. It took talent.

Now, a criminal can automate a lot of it.

They can generate messages, scripts, documents, and voices far faster than before.

That means the number of attempts goes up.

And when attempts go up, the number of victims goes up even if the success rate stays the same.

2) The internet is full of "free material" about us

People don't realize how much of their voice and personality is already public:

- Videos

- Voice notes

- Interviews

- Social media

- Resume

- Podcasts

- Voicemails

Even inside a company, attackers don't need to guess the culture; they can observe it.

They can learn:

- How your CEO writes emails

- Which employees handle payments

- Which vendor names sound familiar

- Which projects are real

The better the story, the easier it is to sell.

3) Work is faster, and everyone is overloaded

Modern work conditions make deception easier:

- People work across too many apps

- Approvals happen quickly

- Teams are distributed

- Verification feels "slow."

- Everyone is multitasking

Attackers aren't just hacking systems. They're hacking *speed*.

The Truth Most People Don't Like Hearing

Here it is:

Your brain is not built for this.

Humans evolved to trust face-to-face signals:

- Body language

- Physical presence

- Familiar voices

- Real-world context

We didn't evolve to question whether our boss's voice is real.

So, we have to adapt on purpose.

Not by becoming paranoid.

By becoming disciplined.

That's what professionalism looks like in the modern cyber world:

Calm systems that don't depend on perfect humans.

The First Rule of the New Era

In the old era, a good rule was:

"Don't click suspicious links."

In the new era, the rule becomes:

"Don't act on urgency without verification."

Because urgency is the weapon.

The scammer doesn't want you to think.

They want you to move.

So your defense must be simple and repeatable:

- Slow down

- Verify through a second channel

- Use a known contact method

- Confirm with a second person when money or access is involved

You don't need to win a battle of technology.

You need to win a battle of **process**.

A Quick Reality Check: This Isn't Just a "Person Problem"

A lot of organizations respond to scams by saying:

"We need better training."

Training helps, but training alone is not enough.

Because the attacker only has to succeed once.

A tired employee on a bad day is not a failure of character.

It's a failure of design.

So professional preparation looks like this:

- Payment processes that require confirmation
- Access processes that require verification
- Recovery processes that don't rely on one inbox
- "Two-person rules" for high-risk actions
- Clear steps for "what to do if something feels off."

When those systems exist, the employee isn't alone.

The company doesn't depend on one moment of perfect judgment.

Small Warning Signs People Miss

These are the things that look "normal" until you train yourself to notice them:

- The message pushes secrecy: "Don't tell anyone."
- The message pushes speed: "Do it in the next 10 minutes."

- The message uses authority: "I'm approving this personally."

- The request breaks routine: "Send it to this new account."

- The request bypasses the process: "Don't open a ticket, just do it."

None of these proves it's a scam, but they're signals.

And in cybersecurity, signals matter.

What This Chapter Is Really Saying

Let's say it clearly:

The future of cybersecurity is not just about stopping hackers.

It's about protecting trust in a world where trust is easier to fake.

That's the new reality.

And everything else in this book builds on that.

Because once you understand that trust is the target, you start to see:

- Why cybercrime is organized (not random)

- Why social engineering keeps working

- Why deepfakes will be used for money, access, and control

- Why "smart people" still get tricked

A Simple Starter Playbook (Just 5 Rules)

This is not the full "how to prepare" section yet this is just the foundation.

1. Treat urgency as a red flag, not a reason to rush.

2. Verify money and access requests using a second channel.

3. Have a "safe phrase" in your family and a verification process at work.

4. Don't rely on caller ID or display names. They're easy to fake.

5. When something feels off, pause. The pause is power.

If you do nothing else, do those. They stop a shocking number of modern attacks.

"Because trust is breaking, cybercrime has reorganized itself into a business."

Chapter 2

When trust became exploitable, cybercrime stopped being chaos and became an industry.

(Markets, roles, scaling)

In Chapter 1, we talked about trust how it's breaking, how it's being faked, and how attacks are becoming quieter and more believable.

That shift didn't happen by accident.

When trust became easier to exploit, cybercrime adapted.

And when cybercrime adapted, it stopped looking like chaos and started looking like a business.

if you still picture cybercrime as one person "hacking" a company from a laptop... you're not alone.

A lot of movies, headlines, and even old-school security training taught us the same image:

one attacker, one target, one dramatic break-in.

But that's not how most cybercrime works anymore.

Today, it's closer to business.

And I don't mean that as a metaphor.

I mean it in the real sense:

- There are specialists
- There are products and services
- There are "suppliers" and "buyers."
- There are partnerships
- There are repeat customers
- There's scale and efficiency
- And there's profit driving all of it

Once you understand that, a lot of modern attacks start to make sense.

Not because they become less dangerous, but because they become **predictable**.

Imagine a company that's doing fine.

Not a giant brand. Not famous. Just a normal organization:

- A finance team
- A sales team
- Email
- Shared files
- Cloud tools
- People working fast

One day, someone logs in as an employee.

No alarms go off.

Because the login looks normal.

Correct username. Correct password. Correct location or close enough.

Maybe the attacker even has the second step (the code) because they tricked the employee into approving it.

Now the attacker is "inside."

No broken windows. No loud crash.

Just a quiet entry.

Over the next few days, they poke around:

- Who approves payments
- Where invoices are stored
- Who has access to payroll
- What vendors the company uses
- Which email conversations feel important

Then they send one message clean, professional, polite pretending to be the CEO or a vendor.

And the money moves.

When people finally notice, they say:

"How did someone hack us?"

But the real question is:

Who sold them the keys?

Who coached the scam?

Who built the tools?

Who helped move the money?

Because it usually isn't one person.

It's a chain.

The Big Shift: Cybercrime Has a Supply Chain

A supply chain is just a fancy way of saying:

Different people do different jobs, and they trade what they produce.

That's exactly what cybercrime looks like now.

Here's the simplest version:

1) Find the door

Some criminals don't run ransomware or scams at all; they just scan the internet looking for weak points.

Their job is to locate easy entries:

- Old software

- Exposed remote access

- Weak passwords

- Sloppy setups

- Reused credentials

They don't need to steal money themselves.

They just need to find openings.

2) Sell access

This is where cybercrime starts to feel like a marketplace.

There are criminals whose main product is:

"Access to a company network."

They break in first, then sell that access to someone else.

That buyer might be a ransomware group.

Or a fraud crew.

Or someone is stealing data.

The point is: the first attacker isn't always the one who finishes the attack.

3) Run the attack

Now the "main team" takes over.

This is the group that does the thing you read about:

- extortion
- ransomware
- data theft
- payroll fraud
- account takeovers
- vendor impersonation

These groups are often organized. They know what they're doing.

And they don't waste time on targets that won't pay.

4) Negotiate and pressure

A modern extortion group doesn't just encrypt files and hope.

They pressure.

They threaten.

They push your emotions and your reputation.

Some groups even operate like they have a helpdesk:

- instructions
- deadlines
- "Customer support" tone
- Scripted negotiation tactics

It sounds ridiculous until you realize: it works.

5) Move and clean the money

Finally, there's the problem criminals always have:

How do you turn stolen money into usable money without getting caught?

That's why many cybercrime operations include people who specialize in:

- laundering

- mule accounts

- moving funds

- converting value across systems

Again: roles.

Again: business.

Why This Matters: Scale Changes Everything

When crime becomes a business, two things happen:

Attacks get cheaper

Because the work is divided.

One person doesn't need to be a genius at everything anymore.

They can buy what they need:

- Tools

- Access

- Phishing templates

- Stolen data

- Instructions

- Hosting

- "Support"

Attacks get bigger

Because the process becomes repeatable.

If a scam works once, they run it again.

If it works ten times, they automate it.

If it works at scale, they hire more people or partner with other crews.

This is why the future of cybersecurity isn't "a few big attacks."

It's **constant pressure**.

Because the machine is always running.

The "Gig Economy" of Cybercrime

Here's the part most people don't understand until they see it clearly:

Cybercrime isn't one team doing everything.

It's more like a gig economy.

Different groups offer services, such as:

- "I can get you access."

- "I can write the malware."

- "I can run the phishing campaign."

- "I can negotiate with victims."

- "I can sell the stolen data."

- "I can move the money."

That's why shutting down one group doesn't end the problem.

It helps. It slows things down.

But the demand stays.

And where there's demand, someone else steps in.

Ransomware Didn't Disappear, It Evolved

A lot of people still think ransomware means:

"They lock your files."

Sometimes that still happens.

But now the smarter play is:

Steal first. Threaten second.

Because backups have improved in many places.

So criminals adapted.

Now the threat is often:

- "Pay us, or we leak your data."

- "Pay us, or we contact your customers."

- "Pay us, or we publish proof you were breached."

It becomes less about your computers and more about your reputation.

And reputation pressure makes people act fast, especially leaders.

That's why ransomware is no longer just a technical incident.

It's a business crisis.

A legal crisis.

A human crisis.

Why Small Businesses and Regular People Are Targets

Let's clear this up, because it's important:

Attackers don't choose targets based on fame.

They choose targets based on **opportunity and payout**.

Small businesses are attractive because they often have:

- Fewer defenses

- Fewer backups

- No dedicated security team

- No incident plan

- More panic

- Faster "just pay it" decisions

Individuals are attractive because:

- Identity is valuable

- Accounts can be reused across services

- One stolen login can unlock many doors

- People act emotionally when scared

Cybercrime scales by harvesting volume.

That means the future isn't about "big companies only." It's everyone.

The Most Important Mindset Shift

If you remember one thing from this chapter, remember this:

You are not defending against a person.

You are defending against a system.

A system that:

- Learns from what works
- Repeats successful tactics
- Improves over time
- Shares tools
- Sells access
- Spreads risk across many players

That's why modern cyber defense is less about "one perfect tool" and more about **layers and habits**.

Because business-style crime doesn't stop after one failure.

It tries again somewhere else.

What To Do With This Information (Without Panicking)

This isn't a "full playbook" chapter that comes later.

But there are a few simple truths you can act on right now.

1) Make identity harder to steal

Most modern attacks start with gaining access: a login, a password reset, or a tricked approval. If identity is weak, nothing else matters.

2) Reduce easy entry points

Cybercrime loves easy doors: old systems, exposed services, reused passwords. The easy door is where the supply chain starts.

3) Build "blast walls."

Assume something will go wrong at some point.

Your job is to stop one mistake from becoming a full collapse.

That means:

- Limiting permissions
- Separating important systems
- Requiring confirmation for risky actions
- Not letting one inbox control everything

4) Backups are not just "IT work."

Backups aren't a checkbox.

They're survival.

A business that recovers quickly is harder to extort.

5) Build a calm response plan

Cybercriminals win by rushing you.

A plan slows panic.

Even a simple plan is better than none:

- who decides

- who investigates

- who talks to the bank

- who handles customers

- what gets shut down first

- what gets restored first

The future is not just about preventing attacks.

It's about **being ready when prevention fails.**

Why Chapter 3 Matters

When attacks scale like a business, the human layer becomes

the most efficient entry point

It doesn't only attack technology.

It attacks people.

It attacks attention.

It attacks emotion.

It attacks trust.

That's why in Chapter 3, we're going to talk about something

most cyber books avoid:

Why smart people still fall for them

Chapter 3

Why Smart People Still Get Tricked

(Speed + Emotion + Trust= Mistakes)

Let's get one thing straight before we go any further:

When someone gets scammed, it doesn't mean they're stupid.

It usually means the scam was **designed well**.

And that's what makes this chapter important, because if you

believe scams only work on "careless people," you won't build

the habits and systems that actually protect you.

You'll just tell yourself, *"That won't happen to me."*

That's exactly what scammers want.

A smart person. A normal day. One rushed moment.

Picture a capable employee, someone who's sharp, responsible,

and respected.

They're not new. They're not clueless. They're the kind of

person people trust with real work.

They're also busy.

They're in the middle of five things at once:

- Inbox filling up

- Meetings starting

- Deadlines

- Phone buzzing

- Someone asking for help

Then a message comes in that looks completely normal.

It's from a familiar name.

The tone feels right.

The request is small, just one action.

And it carries a little pressure:

"Hey, I need this done quickly. I'll explain later."

If you've ever worked in a real company, you know how that feels.

Sometimes you don't even think.

You just move.

Because your brain is trying to keep the day moving.

That's the moment scams live in.

Not the dramatic moment.

The *ordinary* moment.

Why scams work: they don't attack intelligence they attack autopilot

Most of life runs on autopilot.

Not because we're lazy, but because we have to be.

If your brain had to deeply analyze every email, every message, every call, every request, you'd never get anything done.

So your brain uses shortcuts:

- "That looks like my boss."
- "That sounds like the bank."
- "That's a vendor we use."
- "This is how they normally talk.
- "This is probably fine."

Autopilot is how we survive busy lives.

But autopilot is also where criminals hunt.

Because if they can make something feel familiar, you stop analyzing it.

And once you stop analyzing it, you start obeying it.

The three weapons: speed, emotion, and trust

Most modern scams are built around one or more of these.

1) Speed: "Do it now."

Speed is powerful because it shuts down thinking.

The scammer doesn't want you to verify.

They don't want you to ask someone else.

They don't want you to slow down and notice the weird details.

So they create a time trap:

- "I'm in a meeting."

- "This has to happen in the next 10 minutes."

- "We're on a deadline."

- "Don't delay this."

Speed makes you skip steps.

And skipped steps are where scams win.

If you remember one rule from this chapter, remember this:

When a message pushes urgency, that's not a reason to rush.

That's a reason to verify.

2) Emotion: "Feel first, think later."

Emotion is the shortcut to action.

Scammers love emotions that trigger quick movement:

- Fear ("Your account is locked.")

- Pressure ("This is your responsibility.")

- Guilt ("I really need your help.")

- Authority ("Do this because I'm telling you.")

- Excitement ("You've won, act now.")

- Protection ("Your child is in trouble.")

The point isn't to create a huge emotional breakdown.

The point is to create just enough emotion to steal 30 seconds of your judgment.

That's all they need.

3) Trust: "I'm someone you already know."

Trust is the biggest weapon of all.

Because trust isn't just logical.

Trust is emotional.

Trust is related.

Trust is "I've heard this voice a hundred times."

And now we're entering an era where voices, videos, and messages can be faked well enough to borrow trust.

That's why this isn't just a "cybersecurity" problem anymore.

It's a reality problem.

The uncomfortable truth: smart people are the best targets

This is going to sound strange, but it's true:

Smart, competent, helpful people often make *excellent* targets.

Why?

Because smart people:

- Move fast
- Take responsibility
- Don't want to look incompetent

- Don't want to "bother" others

- Solve problems quickly

- Want to be reliable

Scammers love that.

They weaponize professionalism.

They weaponize "being a good employee."

They weaponize "being helpful."

So the goal isn't to become less helpful.

The goal is to become helpful **with safeguards**.

Why "training" isn't enough

A lot of organizations respond to scams by saying:

"We need more awareness training."

Training helps. But training is not a shield.

Because training tries to fix the problem using memory and willpower:

- "Remember the rules."

- "Spot the signs."

- "Be careful."

But scammers aren't betting against your memory.

They're betting against your *human condition*: busy, distracted, tired, rushed.

So the real solution is not just training.

Its **design**.

Design beats willpower.

A well-designed system protects people on their worst day.

A poorly designed system expects perfect judgment every day.

And perfect judgment doesn't exist.

The "Two-Minute Pause" that saves people

Here's a simple habit that prevents a lot of damage:

When money, access, or sensitive data is involved... pause

for two minutes.

Two minutes sounds small.

But it breaks the spell.

It gives your brain time to shift from autopilot to awareness.

In those two minutes, you do one thing:

Verify through a second channel.

Not the same email thread.

Not replying to the same number.

Not clicking the same link.

A second channel means:

- Call the person using a saved number

- Message them in your company directory

- Ask a second approver

- Use the normal process, even if they say "don't."

If the request is real, verification is a small inconvenience.

If the request is fake, verification is a wall.

The "scripts" that make verification feel normal

A lot of people don't verify because they don't want to seem rude.

So give readers simple phrases they can use without drama:

- "Got it. I'm going to call you back on your usual number to confirm."

- "Before I do this, I need to verify through our normal process."

- "No problem send it through the official approval channel."

- "I can help, but I can't bypass verification steps."

- "If this is urgent, verification becomes even more important."

These lines sound professional.

They don't accuse anyone of being a criminal.

They just protect you.

Why deepfakes make this even harder

In the past, people could rely on small clues:

- Bad grammar

- Weird tone

- Awkward phrasing

- Suspicious voicemail quality

Now, those clues are fading.

Messages are cleaner.

Voices can be cloned.

Fake documents look "official."

Fake screenshots look real.

So the new rule isn't "spot the scam."

The new rule is:

Don't rely on your senses as proof.

Rely on verification as proof.

This is a huge shift.

But it's the shift that keeps you safe in the next decade.

What to do now (simple, practical)

This is not the full "playbook" section yet, which comes later in the book.

But these actions belong in your life *today*:

45

For individuals and families

- Create a **family safe word** for emergencies (a word scammers can't guess).

- If you get a scary call: **hang up and call back** on a known number.

- Treat "urgent money requests" as automatic verification situations.

- Don't trust the caller ID alone. **Ever**.

For work and business

- Make "two-person approval" normal for payments and access changes.

- Require verification for **bank detail changes** and **invoice changes**.

- Create a rule: "No financial action based only on email."

- Make it culturally acceptable to say:

  "I'm verifying because it's my job."

The most professional person in the room is the one who *verifies*.

What's coming next?

And the biggest accelerator of everything we just discussed is AI, because it makes scams cheaper, faster, and more believable at scale. That's where we go next.

Chapter 4

AI Scams That Feel Real

(When the lie sounds perfect)

You're going to hear a lot of noise about AI in cybersecurity, half of it hype, half of it panic.

So let's ground this chapter in something simple:

AI doesn't create new human weaknesses. It scales it.

It takes the oldest tricks in the book, impersonation, urgency, pressure, and fake authorit, and makes them faster, cheaper, and more believable.

And that changes what "safe" looks like.

The Message That Would've Worked on Almost Anyone

A manager gets a voice note.

It sounds exactly like the CFO.

Same pace. Same tone. Same little pauses. Even the same habit of clearing his throat before he speaks.

The CFO says he's in the middle of a meeting and can't type.

He says an acquisition payment needs to be confirmed today.

He says the bank is waiting. He says this needs to be quick.

Then he says:

"I'm sending the details now. Please handle it."

A minute later, an email arrives with wiring information. It looks clean. Professional. Normal.

The manager doesn't think, *"This is a scam."*

The manager thinks, *"This is stressful, but it's my job."*

That's the trap.

Because in the AI era, scams don't feel like scams. They feel like work.

What AI Changes in One Sentence

In the old world, criminals had to choose between:

- **personalized** scams (high effort, limited scale)

  or

- **mass** scams (low effort, low believability)

AI removes that tradeoff.

Now, criminals can run scams that are: **Personalized and scalable.**

That's the shift that matters.

The New Scam Factory

Here's the clean truth:

Most scams used to be limited by time.

A person can only write so many convincing emails.

A person can only make so many calls.

A person can only research so many targets.

AI changes the economics.

With AI, a criminal operation can produce:

- Thousands of polished emails that don't sound fake

- Scripts that match a company's tone

- Messages that reference real projects

- Fake "internal" documents that look official

- Responses that sound calm and human

- Endless variations so filters can't easily block them

Not because AI is "evil."

Because it's good at one thing scammers love:

Producing believable language on demand.

Why This Hits Businesses Harder Than People Expect?

Many leaders assume the main threat is employees clicking links.

That's part of it, but AI goes after something more valuable:

Business processes.

AI scams are especially effective in environments where:

- Speed is praised

- People are overloaded

- Approvals happen quickly

- Employees are trained to be helpful

- "Just get it done" is the culture

AI doesn't need to break the firewall if it can break the workflow.

So instead of attacking your network first, criminals increasingly attack:

- Invoicing

- Payroll

- Vendor onboarding

- Bank detail changes

- Password resets

- IT support

- Executive assistants

- Finance teams

These are "trust-heavy" systems.

And trust-heavy systems are where AI impersonation hits hardest.

Deepfakes Aren't the Whole Story

When people hear "AI scam," they think deepfake videos.

Deepfakes are real, and they're growing, but the biggest danger isn't always video.

The biggest danger is something quieter:

AI makes everyday communication "cheaply convincing."

A clean email with the right tone can be more dangerous than a fake video, because email is already normal. Nobody treats email like "evidence." People treat it like work.

That's why AI scams succeed:

They don't need to shock you; they need to blend in.

The Death of "Vibes" as a Security Strategy

For years, many people relied on a feeling:

- "This looks legit."

- "This sounds like her."

- "This seems normal."

In the AI era, that approach breaks.

Because AI is good at producing *vibes.*

It can mimic professionalism.

It can mimic urgency.

It can mimic authority.

It can mimic friendliness.

So here's the new rule:

You can't verify truth by how real it feels.

You verify truth by process.

That's a big shift. And it's the core of professional cybersecurity in the future.

The Four Types of AI Scams You'll See Everywhere

1) AI-written impersonation (Email + Text)

This is the simplest and most common.

AI helps attackers write messages that:

- Sounds like a real coworker
- Match a company's writing style
- Referred to relevant details
- Avoid spelling mistakes and weird grammar
- Keep the tone calm and believable

This kills one of the old defenses people used: "I can tell it's fake because it sounds fake."

Now it won't sound fake.

2) Voice impersonation (Calls + Voice notes)

Voice is powerful because it automatically triggers trust.

When you hear a familiar voice, your brain stops questioning.

Even if you're careful, the voice hits a deep part of human wiring:

Voice = Person.

In the future, that shortcut will become dangerous.

3) Fake proof (Screenshots + documents)

One of the fastest-growing tricks is "proof theater."

Attackers send:

- A screenshot of a "bank confirmation."

- A PDF that looks like an internal approval

- A document that looks like a vendor contract

- A fake ticket number

- A fake invoice chain

Not because the proof is real.

Because proof makes people stop asking questions.

4) Conversation scams (Back-and-forth realism)

Older scams were one-message and done.

AI enables something more effective:

A real conversation.

- You ask a question

- The scammer answers smoothly

- You hesitate

- They reassure you

- You push back

- They adjust

Those back-and-forth increases trust.

It also increases compliance.

Why This Will Keep Getting Worse?

Not forever. But for the next few years, yes.

Because three forces are lining up:

1. **AI tools keep improving** (quality rises)

2. **AI costs keep dropping** (volume rises)

3. **More of life moves through digital channels** (opportunity rises)

That's why the "future of cybersecurity" isn't a single big event.

It's a steady pressure.

Like weather.

How to Prepare Without Becoming Paranoid

Here's the professional approach:

You don't need to "spot every deepfake."

That's not realistic.

Instead, you build **verification habits** and **approval systems** that remain strong even when communication gets faked.

Preparation is not suspicion.

Preparation is structure.

The Practical Defense: Upgrade Your Verification, Not Your Fear

1) Second-channel verification becomes standard

If a request involves:

- Money

- Access

- Sensitive data

- Account changes

- Security settings

...it gets verified on a second channel.

Not "reply to the same email."

A real second channel is:

- Call back using a number from your directory

- Confirm in person

- Confirm through an internal tool that the attacker can't easily fake

- Require a second approver

2) Build "no single point of human failure" rules

Professional organizations stop relying on a single person's judgment for high-risk decisions.

Examples:

- Two-person approval for wire transfers
- Verified vendor change procedure
- Payroll change confirmation
- "No payment based on email only."
- Mandatory delay for new bank account details

These rules aren't about *distrust*; they're about *reality*.

3) Normalize the sentence: "I'm verifying because it's my job."

Scammers win when people feel awkward verifying.

So you make verification culturally normal.

You train people to say it calmly, without apology.

Verification is professionalism now.

4) Separate "communication" from "authorization."

This is a big one.

In many companies, a message becomes an action.

That's risky.

The future-safe model is:

- messages request actions
- **systems authorize actions**

Meaning: even if the message is perfect, it still can't bypass the approval system.

Quick Checklist: AI Scam Readiness

Use this as a "Week 1" readiness list.

For individuals

- Create a family safe word (one word that confirms it's real).
- If you get an urgent call: hang up and call back on a saved number.
- Treat urgent money requests as automatic verification events.
- Don't trust caller ID as proof, **ever**.

For companies/teams

- Two-person approval for payments and access changes.
- A written process for bank detail changes (no exceptions).
- A standard "call-back rule" for urgent executive requests.
- Train staff with realistic scenarios, not just slides.
- Make it safe to slow down: no shame for verifying.

The Real Point of This Chapter

AI is going to make scams more believable.

So the professional move is not "try to detect perfect fakes."

The professional move is:

Design your life and your business so that believable lies don't get automatic power.

That's how you win the next era.

Coming Next

In the next chapter, we move from scams to disruption:

Ransomware and extortion.

Because, as AI makes deception easier, criminals don't just steal.

They pressure. They threaten. They squeeze.

And the organizations that survive won't be the ones with the fanciest tools.

They'll be the ones who can **recover fast and stay calm.**

Chapter 5

Ransomware Becomes "Extortion First"
(Why aren't backups enough anymore)

Ransomware used to be simple.

Not easy. Not harmless. But simple.

The old deal looked like this:

"We locked your files. Pay us, and we'll unlock them."

That was the story most people learned.

So companies responded with a smart plan:

- Keep backups

- Rebuild systems

- Restore operations

- Refuse to pay

And for a while, that worked more often.

So criminals adapted.

Now the modern deal often looks like this:

"We already copied your data.

We can still lock your systems if we want.

But even if you restore everything... We'll leak what we stole."

That's **extortion first**.

And it changes the whole game.

The Attack That Doesn't Start with Chaos

Here's how it happens in real life.

A company is running normally. People are working. Emails are flowing. Customers are ordering. Nothing looks wrong.

Behind the scenes, someone is already inside.

Not crashing systems. Not making noise.

Just watching.

They learn:

- Who has admin access.
- Where customer records live.
- Where contracts live.
- Where payroll lives.
- Which systems matter most?
- What kind of business is this?
- What would hurt the most if it became public?

Then, when they're ready, they do one of two things:

Option A: They encrypt and lock everything

The "classic" ransomware moves.

Option B: They skip encryption and go straight to pressure

Because the real weapon is not the lock.

It's the leverage.

It's the threat of exposure.

It's the fear of reputational damage.

It's the fear of lawsuits, regulators, customers leaving, and competitors gaining an advantage.

And that's why ransomware isn't just an IT problem anymore.

It's a **business survival** problem.

Why "Extortion First" Works So Well

Extortion-first ransomware works because it attacks something humans care about deeply:

Control of your story.

Most companies can survive an internal mess.

What they can't survive is losing control publicly.

Attackers know that.

So they don't just threaten your systems.

They threaten your trust.

They threaten your name.

They threaten your relationships.

And that's often more powerful than encryption.

Because encryption creates one problem:

Downtime.

Data theft creates multiple problems:

- Downtime

- Reputation damage

- Legal exposure

- Customer loss

- Partner loss

- Employee fear

- Long-term cleanup

So even if you have backups, you still feel trapped.

That's the new pressure.

The New Ransomware Playbook

Modern ransomware groups don't rely on a single tactic.

They stack tactics.

Here's what "extortion first" can include:

1) Data theft

They copy what hurts most:

- Customer data

- Employee records

- Contracts

- Invoices

- Private communications

- Internal documents

2) Proof

They send you screenshots and samples like:

- "Here's your payroll file."

- "Here are your contracts."

- "Here are your customer IDs."

This is psychological.

They're saying:

"We're not bluffing."

3) The threat

Now they offer the deal:

- Pay, and we won't leak it

- Pay, and we'll "delete it" (they claim)

- Pay, or we publish it

- Pay, or we contact your customers

4) Optional escalation: encryption

If they want more pressure, they lock systems too.

5) Optional escalation: DDoS and harassment

Some groups add more pain:

- Knocking your website offline

- Emailing your customers

- Threatening executives

- Contacting media

This is why it's called extortion, not "ransomware."

Ransomware is just one tool in the extortion toolbox.

Why This Is Getting Worse (Even If Security Improves)

You'd think better security would reduce this problem.

And in some ways, it does.

But the business model keeps ransomware alive.

Here's why:

1) Crime follows profit

Extortion is profitable.

If something makes money, it attracts more criminals.

If it attracts more criminals, it evolves faster.

2) Companies are dependent on speed

Modern businesses can't afford long downtime.

Even a small outage can destroy a small business.

Attackers know exactly where the pressure is.

3) Data is everywhere

Most organizations have way more sensitive data than they think.

Data spreads into:

- Email

- Shared drives

- Cloud apps

- Personal devices

- Old systems nobody remembers

- Third-party vendors

That creates more leverage opportunities.

4) The human factor remains

Ransomware doesn't need perfect exploit.

It often begins with:

- One stolen password

- One successful phishing message

- One exposed system

- One careless approval

And attackers only need one opening.

The Hard Truth: Paying Doesn't Solve It Cleanly

People ask:

"Should we pay?"

That decision is complicated, and it depends on the situation.

But here's what I want you to understand clearly:

Paying is not a magic undo button.

Paying can bring risks too:

- You might not get a full recovery

- You might be targeted again

- You might still face data leaks

- You might be marked as "willing to pay"

- You might still have legal and regulatory issues

And there's a deeper truth:

You can't buy back time.

Even if systems come back, the disruption has already happened.

So, the smartest goal is not *"never pay."*

The smartest goal is:

Build an organization that doesn't get cornered.

Because ransomware is really about cornering you:

Emotionally, operationally, reputationally.

Preparation is how you avoid being trapped.

The Three Clocks That Decide Everything

When ransomware hits, three clocks start ticking.

Clock 1: The attacker's clock

They're pressuring you, setting deadlines, escalating threats.

Clock 2: Your business clock

How long can you operate like this?

How long until customers notice?

How long until revenue collapses?

Clock 3: The exposure clock

How long until this becomes public?

How long until data leaks?

How long until partners find out?

The best defenders win by controlling these clocks.

Not by being perfect.

By being ready.

What "Ready" Actually Looks Like

Many companies say they're prepared because they have "security tools."

That's not the same thing as being ready.

Ransomware readiness is a mix of:

- Technology

- Process

- Leadership

- Practice

- Communication

It's not just IT.

It's the whole organization.

So here's the real preparation list, written in human terms, not like a policy document.

The Extortion-First Preparation Plan

1) Backups that can actually save you

Not "we think we have backups."

Backups that are:

- **Regular**

- **Protected from tampering**

- **Tested**

- **Fast enough to restore your business**

The number one lie in ransomware response is:

"We have backups."

Then the day comes, and they discover:

- Backups failed months ago

- Backups are also encrypted

- Restore takes weeks

- Nobody knows the order of recovery

Backups are only real if you test restoring.

If you don't test restore, you don't have backups.

You have hope.

2) Reduce the "blast radius."

A ransomware event should not be able to take everything at once.

So the goal is simple:

One compromise should not equal total collapse.

That means:

- Limit access so one account can't reach everything
- Separate critical systems
- Protect admin accounts more than regular accounts
- Avoid "everyone has access to everything" culture

This is one of the most professional moves a company can make.

It's not about paranoia, it's about containing reality.

3) Protect identity like it's your front door

Most ransomware events still begin with access.

That's why identity is the battlefield.

Make it harder to steal access:

- Strong login protection
- Extra verification where it matters
- Fast response to unusual logins
- Careful control over who can approve what

If attackers can't get reliable access, extortion becomes harder.

4) Have an incident plan before you need it

When ransomware hits, the biggest enemy is chaos.

People panic.

People argue.

People guess.

People delay.

A plan turns panic into action.

A simple plan answers:

- Who is in charge of decisions?
- Who talks to the bank, legal, insurance, and leadership?
- Who talks to customers if needed?
- What gets shut down first?
- What systems get restored first?
- How do we preserve evidence without making it worse?

You don't need a 50-page document.

You need a **clear, practiced plan.**

5) Practice like it's a fire drill

Most organizations do fire drills.

But cyber incidents often get treated like "we'll figure it out."

That's a mistake.

Do a simple tabletop exercise:

- "What if our file share is locked?"

- "What if customer data is stolen?"

- "What if payroll is hit?"

- "What if attackers contact customers?"

When you rehearse it once, your response improves

dramatically.

Practice makes your team calmer.

Calm teams make better decisions.

6) Reduce the leverage in the first place

Extortion works when attackers steal something valuable.

So ask a professional question:

"If someone stole our data tomorrow, what would hurt us

most?"

Then reduce it:

- keep less sensitive data where it doesn't need to be

- archive properly

- limit who can access it

- avoid storing secrets in random shared folders

- know where your "crown jewels" are

You can't protect what you can't find.

"If You Only Do Five Things" Checklist

If you're a small business or you want a simple starting point, this is it:

1. **Test your backups** (restores something, don't just assume)

2. **Protect money actions with two-person approval**

3. **Lock down admin access and reduce "everyone can access everything."**

4. **Write a one-page incident plan** (who does what)

5. **Create a calm verification culture** (no shame in slowing down)

That one checklist prevents a shocking amount of damage.

Closing: Why This Chapter Matters for the Future

Ransomware isn't going away soon.

Not because defenders are doing nothing.

But because extortion is profitable, the tactics keep evolving.

And in the next era, the danger won't just be "your files are locked."

It will be:

*   Your data is copied

*   Your reputation is threatened

- Your business is pressured

- Your customers are pulled into it

So the future-proof strategy isn't just prevention.

It's resilience.

It's recovery.

It's not getting cornered.

Coming Next

In the next chapter, we step deeper into the real battlefield

behind scams and extortion:

Identity becomes the main battleground.

Because once attackers can steal or imitate identity, they don't

need to "hack" the old way, they walk through the front door.

Chapter 6

Identity Is the New Perimeter

(Why attackers don't "hack in" anymore,

they log in)

Let's start with a situation that doesn't feel like a cyberattack.

You're working. Your phone buzzes with a notification:

"Approve sign-in?"

Yes / No

You didn't try to sign in.

You stare at it for a second. Then it buzzes again.

And again.

Now you're annoyed. You're in the middle of something. You think:

"Maybe it's a system glitch."

"Maybe IT is doing something."

"Maybe I accidentally triggered it."

You hit **Yes** just to make it stop.

And in that moment, without downloading a virus or clicking a shady link, you just opened the door.

Not into a computer.

Into your identity.

And once someone has your identity, they don't need to fight firewalls.

They don't need to break windows.

They just walk in as if they belong there.

That's the new perimeter.

The Old World Had Walls. The New World Has Logins.

For years, cybersecurity was built like a building:

- You had a "network." It had a boundary
- You defended the boundary
- Inside the boundary felt safer than outside

That model made sense when most work happened in one place, on one network.

But now your data doesn't live "inside the office" anymore.

It lives in:

- Cloud apps
- Email systems
- Shared documents
- Collaboration tools
- Customer databases

- Payroll platforms

- Finance systems

- SaaS dashboards

- admin portals you access from anywhere

So, the wall doesn't matter the way it used to.

The front door does.

And the front door is almost always:

A login.

That's why identity is the battlefield.

What "Identity" Really Means (In plain language)

When most people hear "identity," they think "username and password."

That's part of it.

But identity is bigger than that. It's the whole system of proof:

- Your password

- Your MFA (the extra code or approval)

- Your password reset process

- Your email account (because it resets everything else)

- Your phone (because it often receives verification codes)

- Your sessions (the "you're already logged in" state)

- Your recovery options

- Your trusted devices

- Your approvals and permissions

So identity is not one lock.

It's a chain of locks.

And attackers don't need to break all of them.

They need the weakest link.

Why Attackers Love Identity Attacks

Because identity attacks are:

Quiet

No alarms. No broken systems. No loud malware.

Fast

If a criminal gets valid credentials, the "hack" is basically done.

Scalable

They can try the same trick against thousands of people.

Hard to argue with

If the system says "correct password," it treats it as a real user.

That's the scary part:

A computer doesn't know the difference between *you* and

someone using your access.

It only knows whether the keys work.

The Three Most Common Ways Identity Gets Stolen

1) They steal your password

Sometimes it's phishing. Sometimes it's a data leak. Sometimes it's reused passwords.

The details change, but the result is the same:

They get a password that works somewhere.

2) They trick you during login

This is the modern sweet spot:

They don't need to break anything. They just need you to approve something you shouldn't.

This includes:

- Fake login pages
- "Verify your account" messages
- Fake IT support calls
- Repeated MFA prompts until you accept just to stop them

3) They steal the recovery path

This is one of the most overlooked dangers.

If someone gets control of:

- Your email or your phone number

…they can reset passwords for other accounts.

That means your *recovery* becomes their *entry point.*

That's why "protect your email" is no longer optional.

Your email is not just an email; it's the master key to your whole digital life.

The Real Problem: Most Organizations Still Treat Identity Like a Side Topic

Many businesses spend serious money on security tools… but treat identity like an IT setting:

- "We turned on MFA, we're good."

- "We have strong passwords, we're good."

- "We use a login portal, we're good."

But identity security isn't one feature; it's an entire discipline.

Because attackers don't need to beat your best defenses.

They only need to beat:

- One weak user

- One weak process

- One weak recovery path

- One forgotten admin account

- One overpowered permission

- One integration nobody remembers setting up

Identity security is about eliminating "Easy Yes."

The Most Dangerous Sentence in Cybersecurity

Here it is:

"It's probably fine."

Identity attacks succeed because they look normal.

A login from a new location might be normal.

A password reset might be normal.

An MFA prompt might be normal.

So people treat abnormal moments as harmless.

Attackers love that.

They hide inside **"probably."**

The Professional Fix: Stop Treating Logins Like Trust

This is the mindset shift:

A login is not proof of trust.

A login is only proof that a credential worked.

So modern security is moving toward a better idea:

Continuous verification.

That means:

- Not just "who are you?"

- but also "Does this behavior match you?"

- "Is this device normal?"

- "Is this request risky?"

- "Is this happening at a weird time?"

- "Should we ask for extra proof?"

You don't have to be paranoid.

What You Can Do About It

Let's get practical.

This chapter isn't here to scare you, it's here to make identity harder to steal.

For individuals

1) Use a password manager (and stop reusing passwords)

Reused passwords are one of the oldest ways criminals get in.

A password manager isn't "extra."

It's how normal people survive modern security.

2) Turn on multi-factor authentication everywhere you can

Yes, it's annoying sometimes.

But it's one of the few defenses that actually blocks a lot of real attacks.

3) Protect your email like it's your bank account

Because it basically is.

If someone takes your email, they can reset everything else.

So:

- Strong password

- MFA

- Recovery options locked down

- Don't let random apps connect to it

4) Treat unexpected MFA prompts like a warning

If you get a login approval you didn't trigger:

- Hit **No**

- Change your password

- Check account activity

- Don't "approve to make it stop."

Because that "stop" is the attacker trying to push you into autopilot.

5) Upgrade your "verification habits" for urgent requests

If someone asks for money, access, or sensitive info:

- Verify on a second channel

- Call back on a known number

- Don't trust caller ID **ever**

- Don't trust "it sounds like them."

In the AI era, your senses are not enough.

For companies and teams

This is where identity becomes a leadership issue, not just an IT issue.

1) Make high-risk actions require a second human

Especially for:

- Wire transfers
- Payroll changes
- Vendor bank changes
- Account permission changes
- Password resets for executives
- New device approvals

No one person should be able to take irreversible actions alone, especially not under pressure.

2) Separate "normal accounts" from "power accounts."

In plain language:

People shouldn't browse email and click links using the same account that can change critical systems.

If a powerful account gets tricked once, the blast radius is massive.

Professional security reduces that blast radius on purpose.

3) Fix the reset process (because it's where attackers love to play)

A weak password reset process is basically an open door with a polite sign on it.

Make resets require verification:

- Call-back rules

- Manager approval for sensitive roles

- Strong identity checks

- Clear steps that don't get bypassed just because someone sounds urgent

4) Remove "forever access."

The longer someone keeps access to something they don't need, the more likely it is to be abused.

Professional organizations make access temporary when possible:

- Access for a task

- Access for a time period

- Access with approval

This reduces long-term risk without killing productivity.

5) Watch for "impossible behavior."

You don't need spy-level monitoring.

You need basic awareness, like:

- Logins from unusual places

- Sign-ins at unusual times

- Sudden changes in permissions

- Suspicious download activity

- Multiple failed login attempts

It's not about spying on employees.

It's about noticing when identity stops behaving like the real person.

The Hidden Identity Problem: "Non-Human" Accounts

Here's something many people miss:

Not every "identity" is a person.

Companies have identities that belong to:

- Integrations

- Automated services

- Apps connected to apps

- Background processes that move data

These identities often have:

- Powerful access

- Weak oversight

- Long lifetimes

- No MFA

- No one is watching them daily

So professional move is simple:

If something has access, it needs ownership and review.

"Who created this?"

"Why does it still exist?"

"Does it need this much permission?"

"What happens if it gets stolen?"

If nobody can answer those questions, that's a risk.

The Identity Rules That Work in Real Life

If you want a simple set of rules that keeps you safe in the next decade, use these:

1. **No urgent money movement without second-channel verification.**
2. **No account reset without identity confirmation.**
3. **No powerful access on the same account that reads email all day.**
4. **No "forever permissions" without review.**
5. **No ignoring unexpected MFA prompts; treat them as an active attack.**

These rules aren't technical.

They're behavioral and procedural.

And they work because they attack the root problem:

87

identity is being targeted.

Closing: Why This Matters for the Future

The future of cybersecurity isn't just about better malware.

It's a more believable impersonation.

More stolen logins.

More identity-based entry.

Because once criminals can reliably steal an identity, they don't need to smash doors.

They just use keys.

And the modern world is built on keys.

Coming Next

In the next chapter, we take this one step further:

Even if your identity is strong, you still connect to other people and other companies.

Vendors. Partners. Contractors. Software providers.

And that's where the quiet danger lives:

Third parties and supply chains.

Because sometimes the most straightforward way into your world... is through someone you trusted to connect to it.

Chapter 7

The Quiet Threat:

Vendors and Supply Chains

The easiest way into you is through someone

you trust.

Most companies don't get breached because they're careless.

They get breached because they're connected.

That's the part people hate hearing, because connection is the whole point of modern business.

You don't run a company alone anymore. You run it with:

- Cloud services

- Payment providers

- Accountants

- Marketing tools

- Contractors

- IT support

- Software updates

- Plugins

- HR platforms

- Email tools

- Shipping tools

- Vendors you barely think about

And every connection is a door.

Some doors are strong.

Some are weak.

And attackers don't care which door is "fair."

They care which door opens.

The breach that starts with "not us."

Here's how it happens.

A company does a lot of things right:

- Strong passwords

- MFA

- Security training

- Backups

- Careful finance approvals

They feel responsible. They feel prepared.

Then one morning, something weird shows up:

- A login from a trusted service

- A connection from a vendor tool

- A "normal" integration doing something abnormal

- Files being accessed by a third-party account

- Data leaving in a way nobody recognizes

The internal team looks around and says:

"Nothing's wrong on our side."

And that's true.

The attacker didn't come through *your* front door.

They came through a side door you gave to someone else.

This is the supply chain problem in one sentence:

You can secure yourself and still be exposed through your relationships.

That's not paranoia. That's modern reality.

What "supply chain attack" really means (in plain words)

When people hear "supply chain," they think of factories and shipping.

In cybersecurity, it means something simpler:

The attacker uses a trusted third party as a shortcut.

Instead of fighting your defenses, they:

- Compromise a vendor you trust
- Compromise an update mechanism you rely on
- Compromise a service provider with access to many customers
- Compromise a tool connected to your systems

Then they ride that trust straight into you.

Because trust is powerful.

And trust scales.

Why is this getting worse?

Supply chain attacks are rising because they are **efficient**.

1) One vendor can unlock hundreds of targets

If a criminal compromises a tool used by 500 companies, they don't have to attack 500 companies.

They attack one vendor… and inherit access.

That's a dream scenario for cybercrime: high impact, low effort.

2) Modern business is built on integration

Apps talk to apps.

Systems connect through APIs.

Automation runs in the background.

And because it's "normal," people don't monitor it as closely as they monitor employees.

That's where attackers hide.

3) Everyone wants speed

Companies connect tools quickly because it saves time.

But speed creates shortcuts.

Shortcuts create weak points.

And weak points create incidents.

The most dangerous phrase in vendor security

Here it is:

"They're a trusted vendor."

That phrase is not security control.

It's an assumption.

And assumptions get exploited.

A vendor can be reputable and still get breached.

A vendor can be careful and still make a mistake.

A vendor can have good security and still have one weak employee, one exposed system, one overlooked integration.

So the professional mindset is:

Trust vendors operationally, but verify them structurally.

Meaning: you can use vendors, but you don't give them unlimited power.

Where supply chain risk actually shows up (real-world doors)

Here are the most common "quiet doors" that get abused:

1) Vendor accounts with too much access

A contractor gets admin access "temporarily" ... and it never gets removed.

A service provider has a support account that can reach everything.

A vendor tool has permission to read all files "because it was easier."

That's how good intentions become big blast radiuses.

2) Software updates and plugins

Updates are supposed to protect you.

But if an attacker compromises the update path, they can deliver the attack wearing the uniform of trust.

The scary part is that people install updates *because they're being responsible.*

3) Shared credentials and weak vendor hygiene

Sometimes vendors reuse passwords internally.

Sometimes they don't enforce MFA.

Sometimes their staff gets phished.

You don't see that from the outside. You only see the results.

4) "Invisible" tools

Browser extensions. Add-ons. Automation scripts. Small plug-ins.

These don't feel like vendors, but they are.

And they often get more trust than they deserve.

Professional move: reduce the blast radius

You can't eliminate third parties.

But you *can* reduce what happens if one gets compromised.

This is where professional cybersecurity looks like engineering, not fear.

The goal is simple:

A vendor should never have the power to ruin your entire business in one step.

So your strategy becomes:

- Limit access

- Limit time

- Limit permission

- Monitor behavior

- Require verification for sensitive actions

How to prepare (without drowning in paperwork)

Let's keep it real.

Most people hate vendor risk management because they think it means 50-page questionnaires.

That's not where you should start.

Start with habits that actually change risk.

For businesses: 8 practical protections that work

1) Know who your vendors are

This sounds basic, but it's not.

Many organizations don't even have a clean list of:

- SaaS tools

- Integrations

- Plugins

- Contractors

- Service providers

If you can't list them, you can't manage them.

2) Classify vendors by impact

Not all vendors are equal.

A coffee supplier is not the same as your payroll provider.

A professional approach is simple:

- **High impact:** payroll, finance, identity, email, IT support

- **Medium impact:** marketing tools, CRMs, analytics

- **Low impact:** tools with no sensitive access

Start with high-impact vendors first.

3) Enforce MFA for vendor access

If a vendor can log into your systems, their access should require MFA.

Non-negotiable.

4) Give vendors the *least* access they need

Don't give "admin" because it's convenient.

Give access like you're lending your house key:

- Only to the rooms they need

- Only for the time they need

- Only with visibility

5) Remove access when the job ends

This is where many businesses fail.

Access stays forever.

A professional habit is:

- Review vendor access regularly

- Remove unused accounts

- Remove old contractor permissions

- Remove "temporary" access that became permanent

6) Monitor vendor activity like you monitor employee activity

If a vendor account suddenly downloads massive data at midnight, you want that flagged.

You don't need spy-level surveillance.

You just need basic awareness:

- Unusual logins
- Unusual downloads
- Unusual permission changes
- Unusual data movement

7) Separate systems so one breach doesn't spread everywhere

If a vendor connects to one system, that connection should not automatically lead to everything else.

Segmentation is not a fancy concept.

It's just putting doors inside your house.

If one room is compromised, the whole house shouldn't be.

8) Require verification for high-risk actions

If a vendor requests:

- Changing bank details
- Changing payroll info
- Resetting executive accounts
- Changing access permissions

…that should require a verification step and a second approval.

This is how you prevent "vendor compromise" from becoming "company disaster."

For individuals: yes, you have a supply chain too

People forget this, but it matters.

Your personal life has vendors too:

- Phone carrier
- Email provider
- Banking apps
- Cloud storage
- Password manager
- Smart home devices
- Browser extensions
- Apps connected to other apps

The same principles apply:

1) Use fewer apps that have deep permissions

2) Remove apps you don't use

3) Don't install random extensions

4) Protect your email and phone number like a vault

5) Turn on MFA wherever possible

It's the same story on a different scale.

Red flags that should make you slow down

When it comes to vendors and supply chains, these are the warning signs that matter:

- "We need admin access to make it work."

- "We can't use MFA because it's inconvenient."

- "Just share a login, and we'll handle it."

- "Don't worry about the process, we do this all the time."

- "We need full access, otherwise we can't support you."

Professional vendors can work with professional boundaries.

If a vendor refuses basic safeguards, that's not a partnership.

That's risk.

What this chapter is really teaching

Supply chain attacks are not about "bad vendors."

They're about modern dependency.

And dependencies must be managed intentionally.

So, the future-proof posture is:

You don't just secure your systems.

You secure your connections.

That's the new reality.

Coming Next

In the next chapter, we zoom out.

Because vendors are just one part of a bigger problem:

the attack surface keeps growing.

More devices. More apps. More connections. More "doors."

That's Chapter 8: **Everything Is Connected Now.**

Chapter 8

Everything Is Connected Now
The Attack Surface Explosion

It begins the same way modern business always does:

Someone says, "We need a tool for this."

A new app for invoices.

A new CRM.

A new plugin for the website.

A new file-sharing link for a partner.

A new login for a contractor who "just needs access for a week."

And none of it feels dangerous.

It feels like progress.

It feels like work is getting done.

But every new tool, every new login, every new integration is also a new door. And once you have enough doors, you don't need a genius attacker.

You just need an attacker who tries doors all day long until one opens.

That's what "attack surface" means now.

The attack surface isn't a list anymore. It's a lifestyle.

A few years ago, a company's digital world was smaller:

- Office computers
- A server room
- Maybe a website
- Maybe a VPN

Now your business lives in:

- Cloud apps
- Remote work devices
- Shared drives
- Third-party vendors
- Automation tools
- Admin portals you can reach from anywhere
- Links that can be forwarded in seconds

So, the attack surface isn't a single place you can "secure."

It's a moving ecosystem.

And it grows every time you add something.

The biggest misunderstanding

A lot of people hear "attack surface" and think it means:

- Laptops
- Phones

- Servers

- Wi-Fi

That's part of it, but it's not the whole story anymore.

Today the attack surface includes things most people don't count:

- **Accounts you forgot existed**

- **Apps connected to other apps**

- **Shared links that never expire**

- **Admin permissions that were "temporary."**

- **Devices that don't get updated**

- **Old employees who still have access**

- **Tools teams signed up for without telling anyone**

- **Security settings that were left at defaults**

A story you've seen before (even if you didn't call it "cyber")

A small company hires a contractor.

The contractor needs access to a shared folder and one internal tool. Nothing crazy.

To make it fast, someone says:

"Just give them admin. We'll remove it later."

Later never happens.

A few months pass. The contractor's email gets compromised. Or their laptop gets stolen. Or their password is reused somewhere.

Now an attacker has a trusted account inside a real business without breaking a firewall, without writing malware, without doing anything dramatic.

They just inherit access.

And because the account looks legitimate, the activity blends in longer than it should.

That's the attack surface problem in real life:

The danger doesn't always look like danger.

Sometimes it looks like normal operations.

Why the attack surface keeps growing

1) Convenience beats caution (by default)

People choose the fastest path unless something forces them to slow down.

That's not a flaw in people. That's how busy life works.

So "quick access" becomes permanent access.

"Temporary links" become links forever.

"Just this once" becomes routine.

2) Business runs on connections now

Modern companies are built on integrations:

- Finance connects to banking

- HR connects to payroll

- Sales connects to email

- Email connects to password resets

- Password resets connect to everything

So one compromised point can become a chain reaction.

3) No one owns the full map

Most organizations don't have a clear picture of:

- Every app used

- Every admin account

- Every integration

- Every vendor with access

- Every place sensitive data lives

And if you don't know your map, you can't defend it.

The "five surfaces" that matter most

If you want a simple, professional way to think about the

modern attack surface, use these five:

1) People and trust

Scams, impersonation, urgency, approvals, social engineering.

2) Identity

Logins, MFA, password resets, recovery methods, permissions.

3) Devices

Phones, laptops, home Wi-Fi, unpatched systems, BYOD, lost devices.

4) Apps and integrations

SaaS sprawl, plugins, APIs, automation tools, "shadow IT."

5) Data and sharing

Shared links, cloud drives, email attachments, copied files, and old archives.

If you can manage these five, you're doing real cybersecurity, whether you call it that or not.

The professional response: shrink, separate, and standardize

You can't stop the world from connecting.

But you *can* stop the connection from turning into chaos.

1) Shrink what you can

The cheapest security win is removing things you don't need.

- Delete unused accounts

- Removing old employees and contractors

- Uninstall unused apps

- Eliminate "test" systems that became permanent

- Kill old integrations no one remembers

Less surface = fewer doors.

2) Separate what matters most

Not everything should touch everything.

Professional security is about **blast radius**.

If one account gets compromised, the entire business shouldn't fall over.

So, you separate critical areas:

- Finance systems shouldn't be wide open to normal users
- Admin access shouldn't live in the same account you use for email
- Sensitive data shouldn't be accessible "just because it's convenient."

This isn't paranoia. It's engineering.

3) Standardize the rules of trust

The future is full of believable messages and believable requests.

So the only defense that scales is process.

Rules like:

- No money movement from email alone

- No bank detail changes without verification

- No new vendor payments without confirmation

- No "new admin" without a second person approving

- No exceptions just because someone sounds urgent

A practical "attack surface reset" you can do this month

Here's a simple exercise that makes a real difference.

Step 1: Make a one-page inventory

Write down:

- Your top 10 business apps

- Who has admin access to each

- Which three systems matter most to keep running

- Which three systems would hurt most if data leaked

If you can't answer these, you don't have a security problem.

You have a visibility problem.

Step 2: Review permissions like you're cleaning a closet

Ask:

- Who still needs access?

- Who has more access than they should?

- Which contractor accounts are still active?

- Which accounts haven't logged in for months?

- What would happen if this account got stolen?

Step 3: Kill the "forever link" habit

Shared links are convenient and risky.

Make a habit of:

- Expiring links

- Limiting who can access them

- Avoiding open public links for sensitive data

- Checking what's shared externally

Step 4: Decide what gets "two-person approval"

Pick a short list:

- Payments

- Payroll changes

- Bank detail changes

- Admin creation

- High-risk access resets

If it can bankrupt you or expose you, it shouldn't depend on one person having a perfect day.

Takeaway

The attack surface will keep growing. That's the modern world.

So, the goal isn't to "make it small again."

The goal is to **manage it on purpose**, instead of letting it expand by accident.

And if you do that, you don't need to predict every future threat.

You need to make sure the doors stay under control.

Coming next

Now that we've talked about the expanding surface, the next question is:

What happens when the locks themselves start to age out?

That takes us into a quieter but massive shift:

Encryption, and the coming pressure of post-quantum security.

Chapter 9

The Encryption Clock Is Ticking

Why some secrets need to stay safe for years

Most people think encryption is permanent.

Like putting something in a safe and burying it.

But encryption is more like a lock.

And locks don't last forever.

They last until someone builds better lockpicks.

That's what's happening now.

The "time machine" problem

Imagine someone steals a box from your office.

But the box is locked.

You feel relieved because it's encrypted.

Now imagine they don't need to open it today.

They can store it.

They can wait.

And five years from now, ten years from now, when technology is stronger, they will open it then.

That's the danger behind a phrase security people have used for years:

"Harvest now, decrypt later."

The threat isn't just someone breaking your encryption today.

The threat is someone collecting encrypted data today because they believe they can break it later.

And for certain kinds of information, "later" still matters:

- Government data

- Healthcare records

- Financial data

- Legal documents

- Trade secrets

- Personal identity information

- Anything that stays valuable for a long time

Why this matters even if you're not a scientist

You don't need to understand quantum physics to understand the risk.

Here's the plain version:

Some of the math that protects today's internet is based on problems that are hard for regular computers.

Quantum computing aims to solve some of those "hard" problems much faster.

That means organizations need to prepare for a future in which some older forms of encryption are no longer secure.

Not everything breaks the same way, and not everything breaks at once.

But the transition is slow, and that's the problem.

Because even if the "break" is years away, migration takes years too.

Why changing encryption is so hard

People hear "upgrade encryption" and think it's like updating an app.

It isn't.

Encryption is buried inside:

- Websites

- VPNs

- Mobile apps

- Device firmware

- Internal systems

- Third-party services

- Old hardware that still runs critical processes

- Customer integrations you can't easily change

And many organizations don't even know which crypto they're using.

So, they can't upgrade what they can't find.

What preparation looks like (in human terms)

This doesn't need panic. It needs planning.

1) Identify what needs to stay secret for a long time

Ask a simple question:

If this data became readable five years from now, would it hurt us?

If yes, it belongs in your "long-life secrets" category.

2) Build a crypto inventory (even a simple one)

You don't need to start with perfection.

Start with visibility:

- Where is encryption used?
- Which systems rely on older encryption methods?
- Which vendors handle your sensitive data?
- Where do you store old archives?

3) Demand "crypto agility" from vendors

When you evaluate tools and providers, ask:

- Can you update cryptography without rebuilding everything?
- Do you have a plan to support newer standards as they become common?
- How will you handle long-lived encrypted data?

You don't need a vendor to predict the future perfectly.

You need them to prove they can adapt.

4) Stop "custom crypto" behavior

If a system uses home-built encryption, weird proprietary designs, or outdated defaults, that's future debt.

Professional security means using modern, well-supported cryptography that can evolve.

5) Treat this like a migration project, not a patch

The companies that handle this well will treat it like:

- Inventory
- Testing
- Staged rollout
- Vendor coordination
- Timelines
- Fallback plans

Not a one-day update.

The takeaway

The future isn't only about stopping scams and ransomware.

It's also about rebuilding the foundations quietly, before they crack.

Encryption is one of those foundations.

And the smartest organizations won't wait for a crisis to begin the transition.

They'll start by getting visibility now.

Coming next

Next, we'll move from "locks" to "architecture":

how modern organizations are shifting toward **Zero Trust** not as a buzzword, but as a survival strategy in a world where identity and access are under constant attack.

Chapter 10

Zero Trust Becomes the Default

Why "inside the network" doesn't mean safe

anymore

There's a moment that happens in a lot of breaches, and it's always the same kind of moment.

An attacker gets a login.

That's it.

Not a movie scene. Not code flying across a screen. Not a dramatic "system hacked" message.

Just... a successful sign-in.

And once that login works, the old security model quietly collapses, because the old model was built on a dangerous assumption:

If you're inside, you're trusted.

That assumption used to be normal. It made sense when people worked in a single building on a single network.

But we don't live in that world anymore.

Now we work from everywhere, with cloud apps, shared links, vendors, and devices that come and go. The "inside" isn't a place. It's a bunch of connections.

So security has to change, too.

That's where Zero Trust comes in.

What does Zero Trust actually means (in human language)

Zero Trust sounds harsh, like it means:

"Trust nobody. Treat everyone like a criminal."

That's not the point.

Zero Trust really means this:

Trust is not automatic anymore.

Trust is checked every time it matters.

It's the difference between:

- **"You're logged in so that you can do anything."**

  and

- **"You're logged in, but we still verify this request based on risk."**

It's not personal.

It's just modern reality.

Why the old model fails in the cloud era

The old model worked like a castle:

- Big walls (firewalls)

- One main gate (VPN)

- Guards watching the gate

- People inside were treated as "safe."

But modern business is not a castle.

It's a city made of glass.

Your data is in SaaS tools. Your employees are remote. Your apps talk to other apps. Your partners connect to you. Your admins log in from everywhere.

So, the question becomes:

How do you protect a city when there is no single wall?

Answer: You stop depending on the wall.

You protect access itself.

Zero Trust is not "one product you buy."

This is important.

A lot of people think Zero Trust is a tool they can purchase and install.

It's not.

Zero Trust is an approach.

It's a way of running security that asks:

- Who is requesting access?

- Where are they trying to reach?

- Are they using a trusted device?

- Is this normal behavior for them?

- Is the request risky?

- Should we require extra verification?

- If they get compromised, how far can they go?

Zero Trust is less about buying something.

It's more about designing your systems so one stolen login

doesn't become a full disaster.

The four ideas that make Zero Trust work

1) Verify explicitly

Don't assume a user is valid just because they have a password.

Require proof:

- Strong login protection

- MFA where it matters

- Smart rules for risky situations

- Careful account recovery

This is about making "logging in" harder for the wrong person.

2) Use least privilege (give only what's needed)

If someone only needs access to one folder, they shouldn't have

access to ten.

Least privilege is how you stop a small compromise from becoming a significant breach.

3) Assume breach

This isn't pessimism.

It's professional thinking.

It means you design your environment with the expectation that:

- Someone will click something
- Someone will get phished
- Someone will reuse a password
- Someone will lose a device
- Someone will approve the wrong prompt

Assume breach, and you build "blast walls."

4) Segment what matters

If an attacker gets into one part of your environment, they shouldn't automatically reach everything else.

Segmentation is just internal doors.

It's the difference between:

- "One key opens the whole building."

  and

- "This key opens one room."

A simple way to picture Zero Trust

Think about a hotel.

Just because you're inside the hotel doesn't mean you can walk into any room.

You still need a keycard, and your keycard doesn't open every door.

That's Zero Trust.

Not "we think you're evil."

Just: "access is controlled by design."

What Zero Trust looks like in real life (not in buzzwords)

Here are the real-world behaviors Zero Trust creates:

- Admin accounts aren't used for reading email all day.

- High-risk actions (money transfers, new admin creation, bank changes) require a second approval.

- Old accounts are removed quickly.

- Vendors don't get unlimited access "because it's easier."

- Devices are checked (patched, healthy, trusted) before they touch sensitive systems.

- Sensitive systems are separated from general systems.

- Suspicious logins trigger extra verification instead of silent access.

That's it.

It's not magic.

It's discipline, done consistently.

The biggest Zero Trust mistake: making it painful

Some companies hear "Zero Trust" and create a nightmare:

- Constant logins

- Endless prompts

- Productivity collapse

- Staff workarounds

That's not Zero Trust.

That's bad design.

Good Zero Trust is smarter:

Make normal work smooth.

Make risky work verified.

So the system stays usable, but attackers hit friction at the exact moments they need speed.

How to Start Zero Trust Without Overcomplicating It

This chapter is not here to dump a 200-step program on you.

Here's a clean way to begin.

Step 1: Lock down identity first

If identity is weak, nothing else matters.

Start with:

- MFA on key accounts (especially admins and finance)
- Clean account recovery rules
- Remove old users and old access
- Separate admin accounts from daily accounts

Step 2: Decide what actions require "two-person approval"

Pick a short list of actions that can ruin your day:

- Wire transfers
- Vendor bank detail changes
- Payroll changes
- Adding new admins
- Resetting executive accounts
- Changing security settings

Then enforce:

- Verification + second approver
- No exceptions because someone is "in a rush."

Step 3: Reduce "everyone can access everything"

This is where breaches spread.

Tighten access to:

- Sensitive folders
- Customer databases
- Finance and HR systems
- Admin dashboards

Least privilege doesn't make you slow.

It makes you survivable.

Step 4: Put your "crown jewels" behind extra doors

Identify the 2–3 systems that would be most critical if compromised.

Then protect them with extra steps:

- Stronger authentication
- Stricter device rules
- Tighter permissions
- Better monitoring

Step 5: Create a culture where verification is professional

Your people need to feel safe saying:

"I'm going to verify this."

The future is full of believable messages and believable voices.

Verification is not disrespect.

Verification is the new professionalism.

Zero Trust for individuals (yes, it applies)

You might not run a company, but you still have "systems":

- email

- banking

- cloud storage

- social accounts

- phone number (recovery key)

- password resets

A "personal Zero Trust" mindset looks like:

- protect your email like a vault

- use MFA on major accounts

- don't approve prompts you didn't trigger

- verify urgent money requests through a second channel

- reduce connected apps you don't use

Same idea: don't let one stolen login ruin everything.

The takeaway

Zero Trust is not a trend.

It's what happens when the world accepts a hard truth:

"Inside" is no longer safe by default.

Access must be earned continuously.

And when you build systems this way, you don't need perfect

people.

You need consistent rules.

You need clean boundaries.

You need a design that holds up when someone has a bad day.

Coming next

In Chapter 11, we'll shift from access to construction because the future belongs to organizations that build security into the way they develop software, choose vendors, and ship updates. In other words:

security-by-design, or pain later.

Chapter 11
Security-by-Design or pain later
Why "we'll patch it later" is no longer a
plan

Let me start with a story that's painfully common now.

A company doesn't get hacked because someone "broke in"

through the front door.

They get hacked because they *installed something they trusted.*

A tool update.

A plugin.

A library inside an app.

A vendor connection that seemed harmless.

Nobody clicked a weird link.

Nobody downloaded "malware" on purpose.

They just did normal business.

And the attacker used that normal business behavior as the

entry point.

That's where cybersecurity is heading:

The future isn't only about defending systems.

It's about building systems that are harder to betray.

The Shift Nobody Wants to Pay For (But Everyone Eventually Pays For)

For years, a lot of security was reactive:

- Build fast

- Ship the product

- Deal with security "later."

- Patch when something breaks

- Respond when there's an incident

That approach worked *just enough* when the world was simpler.

But today, it breaks down for one reason:

Modern software isn't one thing anymore. It's a stack of other things.

Your app uses:

- Open-source components

- Third-party services

- Cloud infrastructure

- Integrations

- APIs

- Build pipelines

- Authentication systems

- Updates pushed constantly

So, if any layer is weak, the whole structure inherits that weakness.

You don't just need a strong front door.

You need strong materials.

Security-by-Design: The Simple Meaning

"Security-by-design" sounds fancy, but the idea is plain:

Don't bolt security onto the outside.

Bake it into how you build, ship, and operate.

Security-by-design is not perfection. It's not "we'll never be breached."

It's something more realistic and more powerful:

It's reducing avoidable mistakes before attackers can profit from them.

Why Security Has to Move Left

There's a phrase in tech: "shift left."

Ignore the buzzword. Here's what it actually means:

Fix problems earlier.

Because the later you fix security problems, the more expensive they become:

- Expensive to patch
- Expensive to explain

- Expensive to recover

- Expensive to rebuild trust

If you fix a security flaw during design, it's a meeting and a decision.

If you fix it after release, it's:

- Emergency work

- Downtime

- Customer anxiety

- Legal questions

- Reputation damage

The future of cybersecurity is going to punish "we'll fix it later" thinking harder and harder.

The Real Supply Chain: Your Software Has Ingredients

Think about food for a second.

If you buy a packaged meal, you expect an ingredient list.

If there's no ingredient list, you feel uneasy, because you don't know what you're eating.

Software is the same now.

Most software is built from many ingredients:

- Libraries

- Frameworks

- Dependencies

- Components from other teams

- Components from other companies

So a professional organization should be able to answer:

- **What is in our software?**

- **Where did it come from?**

- **Who maintains it?**

- **How do we update it safely?**

- **How do we know it hasn't been tampered with?**

This is why "software supply chain security" has become a big deal.

Not because it's trendy.

Because it's unavoidable.

The Quiet Truth: Attackers Love Your Build Process

A lot of companies focus on protecting production servers.

That's good.

But attackers love attacking earlier in the chain, because if they compromise your build pipeline or your update process, they gain something powerful:

They get your trust delivered for free.

If your system pushes an update that "looks official," people install it.

If your app downloads a component that "looks normal," it runs.

If your developer environment pulls a dependency that "seems fine," it becomes part of your product.

This is why modern defense can't be "block the bad guy."

It also has to be:

"Make sure our own process doesn't deliver the bad guy."

What Security-by-Design Looks Like in Real Life

Let's make this practical.

Here are the habits and decisions that separate "professional" security from "hope-based" security.

1) Secure defaults win the future

The easiest system to compromise is the one that ships with unsafe defaults.

Professional systems default to:

- Least privilege
- Strong authentication
- Encryption on by default
- Logging disabled

- Restrictive access until expanded intentionally

Because most harm happens when someone says,

"We'll tighten it later," and later never arrives.

2) Reduce complexity (because complexity creates hiding places)

Complex systems create:

- Forgotten settings

- Unused accounts

- Misconfigurations

- Weird edge cases

Attackers love edge cases.

So one of the most underrated security moves is:

simplify the environment.

Fewer tools. Fewer permissions. Fewer one-off exceptions.

Security-by-design is often boring like that.

But boring security is the kind that survives.

3) Make "proof" part of development

Modern security is shifting toward evidence, not promises.

Instead of saying, "Trust us, it's secure," you can show things like:

- Code review happened

- Security tests run automatically

- High-risk changes required approval

- Dependencies were checked

- Builds were signed

- Releases were traceable

This isn't about impressing people.

It's about being able to answer calmly when something goes wrong:

"What changed, when, and who approved it?"

That calm answer is the difference between control and chaos.

4) Treat secrets like radioactive material

Most breaches get worse because secrets get exposed:

- API keys

- Passwords

- Tokens

- Private certificates

- Admin credentials

Security-by-design means:

- Don't hardcode secrets in code

- Don't store them in random docs

- Don't reuse them forever

- Rotate them

- Limit who can access them

The more your systems depend on secrets, the more carefully you have to manage them.

Because secrets are what attackers steal to become you.

5) Build with containment in mind

This connects directly to Zero Trust.

Even good software will have flaws.

Even careful teams will miss something.

So security-by-design includes this question:

"If something fails, how far can the damage spread?"

Professional design tries to make sure the answer is:

"Not far."

That means:

- Limiting permissions

- Separating services

- Isolating critical data

- Avoiding "one account rules everything."

- Monitoring behavior

Containment is not a "nice to have."

It's how you survive the modern threat landscape.

The Professional Checklist: If You're Building Anything

If you only do 7 things, do these:

1. **Ship secure defaults.**

2. **Know what's in your software (ingredients list).**

3. **Keep dependencies updated and controlled.**

4. **Automate basic security checks in the build process.**

5. **Protect secrets properly (no shortcuts).**

6. **Limit permissions everywhere (least privilege).**

7. **Design for containment (blast radius).**

That's how you stop "one weak component" from becoming "company-ending damage."

Security-by-Design for Non-Software Companies (Yes, You Still Need It)

Even if you don't build software, you still "build systems":

- Payment processes

- Onboarding processes

- Vendor access processes

- Account recovery processes

- Device enrollment processes

- Approval workflows

Security-by-design applies there too:

Don't design your business so that one rushed click can ruin you.

Design it so high-risk actions require:

- Verification
- Approvals
- Visibility
- Logs
- And clear ownership

That is professional cybersecurity in 2026 and beyond.

What This Chapter Is Really Saying

Security-by-design is not a moral ideal.

It's an economic reality.

Because in a world where cybercrime is organized, AI is scaling scams, and identity is the new perimeter...

The companies that win are not the ones that panic best.

They're the ones that build best.

They build systems that:

- Make risky behavior harder
- Make safe behavior the default
- Make failures smaller
- Make response faster

- Make recovery realistic

This is the quiet foundation of "future-ready" cybersecurity.

And it's why the future belongs to builders, not just defenders.

Coming Next

In Chapter 12, we move from organizations to everyday life:

Personal cybersecurity gets weird.

Because the same forces shaping businesses AI impersonation, identity attacks, and constant connectivity are reshaping what it means to stay safe as a normal person.

And the rules are changing.

Chapter 12

Personal Cybersecurity Gets Strange When your face, voice, and "normal life" become part of the attack surface

A few years ago, personal cybersecurity felt like a small set of rules:

- Don't click sketchy links

- Use strong passwords

- Update your phone

- Watch for obvious scams

That world isn't gone. It's just not enough anymore.

Because the next era of cyber doesn't just target your devices.

It targets **your identity**, your **relationships**, your **reputation**, and the little habits that make life feel normal.

And that's where it gets strange.

Not "science fiction" strange.

Everyday strange.

The moment you realize your identity isn't just "information."

Imagine you're trying to reset an account.

It asks for verification.

Maybe it wants:

- Your phone number
- Your email
- A selfie
- A face scan
- A voice confirmation
- "Prove it's you" in some modern way

That used to feel comfortable.

Now there's a new question hiding under it:

"What if someone can imitate me well enough to pass?"

That's not paranoia. That's the direction things are moving in.

Because personal cybersecurity is shifting from:

protecting devices to **protecting proof.**

And proof is exactly what attackers are learning to fake.

Your life has a "cybercrime profile" whether you want one or not

People don't like to think of themselves as a target.

But criminals don't choose targets the way movies do.

They don't only choose "important people."

They choose **easy wins.**

143

And a "win" can mean:

- Taking over an email account

- Stealing a social media account

- Draining a payment app

- Using your identity to open accounts

- Tricking you into sending money

- Using your name to scam someone else

Most of these attacks don't require genius hacking.

They require **good timing** and **believable communication**.

Which AI makes it easier.

The new personal threats aren't louder; they're more realistic

The old scams were often sloppy.

Bad grammar. Weird tone. Suspicious links.

Now the scam is calmer.

It reads like a normal person.

It sounds like someone you know.

It arrives at the worst moment when you're distracted, tired, or emotionally vulnerable.

That's why this chapter matters:

The future of personal security is less about spotting "obvious bad."

It's about building habits that still work when bad looks normal.

The three personal cyber battles that define the next decade

1) Identity becomes a product

Your identity is valuable because it unlocks everything:

- Bank accounts

- Email accounts

- Shopping accounts

- Medical portals

- Tax info

- Credit

- Your reputation

When criminals steal identity, they're not always trying to "be you" socially.

They're trying to **use you** financially.

And the weird part is: it can happen quietly.

You might not notice until you get:

- A bill

- A loan rejection

- A fraud alert

- A message from someone saying, "Did you ask me for money?"

So identity protection becomes personal cybersecurity's main job.

2) "Reality" becomes less reliable

We're heading into a world where:

- Voices can be cloned

- Screenshots can be faked

- "Proof" can be manufactured

- Messages can imitate tone perfectly

So, the old rule "trust your senses" weakens.

This is why family scams are growing: they don't need to hack your bank first.

They hack your trust first.

3) Your home becomes a network, whether you like it or not

Your home is no longer:

- A door

- A lock

- A mailbox

It's now:

- A Wi-Fi network

- Smart devices

- TVs

- Cameras

- Voice assistants

- Phones

- Laptops

- Work devices

And every device is a possible entry point, not always for dramatic hacking, but for tracking, spying, harassment, or account access through saved sessions.

Personal cybersecurity becomes a "home system" problem, not just a "computer" problem.

What makes it "strange": the new tools of trust

This is the part people feel in their gut:

We're starting to use **biometrics** and **behavior** as proof.

- Face

- Fingerprint

- Voice

- Typing patterns

- Location

- Device behavior

That's not inherently bad. It can be safer than passwords.

But it changes the nature of risk:

If a password leaks, you change it.

If your face/voice becomes part of the verification chain, you can't exactly replace it.

That means you have to treat your identity signals with more care.

Not fear.

Your personal "attack surface" in 2026+ (simple version)

If you want to protect yourself without becoming overwhelmed, think in five surfaces:

1. **Your email** (the master key)

2. **Your phone number** (often the recovery key)

3. **Your financial accounts** (the money target)

4. **Your social accounts** (the reputation target)

5. **Your home network** (the always-on background risk)

If you harden those five, you're ahead of most people.

The professional mindset for personal cybersecurity

This is the same philosophy we used in the business chapters:

Don't try to detect every lie.

Build a life where lies don't get automatic power.

That means your personal security should be designed around two ideas:

1) Verification beats vibes

If something involves money, access, or urgency, you don't act on "it feels real."

You act on verification:

- Call back on a known number
- Use a second channel
- Confirm with a second person
- Slow down on purpose

2) One account should not be able to destroy your whole life

The best personal security change is reducing "single points of failure."

If your email controls everything and your email is weak, you're one bad day away from chaos.

So, you strengthen the master keys first.

A practical personal playbook that doesn't feel technical

Here's what I want you to be able to do after reading this chapter:

Not become a security expert.

Just becomes harder to trap.

Step 1: Protect your email like it's your bank

Because it is your bank, in a modern sense.

Your email resets:

- Your passwords

- Your shopping accounts

- Your social accounts

- Your financial logins

- Your cloud storage

So, the most "professional" cybersecurity move is:

- Strong login protection for email

- Strong recovery settings

- No casual sharing of email access

- Pay attention to "new login" alerts

Step 2: Treat your phone number like a security asset

Phone numbers are used for:

- Account recovery

- Verification codes

- "Proof" that it's you

That's why criminals target phone numbers.

So, your phone security matters more than people think.

Step 3: Stop reusing passwords (or move toward passkeys when available)

Reused passwords are still one of the easiest ways people get pulled into account takeovers.

You don't have to be perfect.

You just need a system so you're not relying on memory and repeated patterns.

Step 4: Make a family "verification culture"

This is one of the most important changes for the AI era.

Families need a simple rule:

"Urgent requests get verified."

Not because you don't love each other.

Because love is exactly what scammers exploit.

A small family habit (like a safe phrase, or a required call-back rule) prevents big emotional and financial pain.

Step 5: Reduce your public "attack material"

You don't need to disappear from the internet.

But you should understand the tradeoff:

The more personal information you put out publicly, the easier it is to write a believable story about you.

This doesn't mean "be afraid."

It means "be intentional."

Why "smart people" still lose personally

Because personal cyberattacks hit human pressure points:

- Embarrassment

- Fear

- Urgency

- Politeness

- The desire to help

- The desire to avoid conflict

That's why advice like "just be careful" isn't enough.

We need systems.

We need habits that work on busy days.

In my first book, I talked about how many breaches begin with phishing or social engineering because humans are the weak link when technology is harder to breach.

The personal version of that truth is simple:

If someone can trick you, they don't need to hack you.

What personal cybersecurity looks like in the near future

Here are a few trends you'll feel more and more:

1) "AI guardians" and smart security helpers

More apps will offer automated protection:

- Fraud detection

- Suspicious login warnings

- "This message looks unusual" alerts

- Automated blocking of risky sign-ins

This can help, but there's a rule:

Automation is a seatbelt, not a steering wheel.

Use it, but don't outsource judgment completely.

2) More "proof theater."

You'll see more fake evidence:

- Screenshots

- Receipts

- Confirmation emails

- Fake support tickets

- Fake ID images

It will look official enough to rush you.

So again: verification > vibes.

3) More attacks that use your account to attack others

A stolen social account isn't just about embarrassment.

It's used to scam your friends because your identity gives the

scam instant trust.

That's why personal security protects your community too.

4) More weird "life admin" risks

The future isn't only "someone stole my password."

It's also:

- Someone changed my payout details

- Someone redirected my recovery email

- Someone added a new login method

- Someone turned off notifications

- Someone quietly set up persistence

That's why protecting recovery and settings matters so much.

The simplest "Personal Cybersecurity Stack."

(If you do nothing else, do this)

If you want a clean, professional baseline, aim for this:

1. **One strong method for managing passwords/logins**

2. **Strong protection for your email account**

3. **Strong protection for your phone and phone number**

4. **Two-step verification for your most important accounts**

5. **A family verification habit for urgent money/emergency requests**

154

6. **A monthly "cleanup" habit**: remove old apps, old access, old devices

This isn't flashy.

But it works because it reduces the most common real-world failure points.

The takeaway

Personal cybersecurity is no longer just "don't get a virus."

It's:

- Don't let someone hijack your identity

- Don't let urgency override verification

- Don't let one account control your whole life

- Don't let your home become a blind spot

- Don't assume "real-looking" means real

That's why it feels weird.

Because it forces a new life skill:

protecting trust in a world where trust is easier to fake.

Chapter 13

Future War Stories

Four near-future cyber scenarios (and what they teach us)

Before we finish the book, I want to do something different.

Because most people don't change their behavior in response to advice.

They change behavior through stories that feel real enough to imagine and clear enough to learn from.

So these "war stories" are not predictions carved in stone. They are plausible futures. built from patterns we already see:

- attacks that borrow trust

- attacks that scale through automation

- attacks that hit operations, not just computers

- attacks that don't end when systems come back online

And after each story, I'll show you the lesson the part you can actually use.

War Story 1: The Voice That Approved the Wire

It's 9:12 a.m. on a Wednesday.

A finance manager is juggling three things at once:

- end-of-month invoices

- a supplier delay

- a meeting in 20 minutes

A voice note lands in Teams/WhatsApp/whatever the company uses.

It's the CFO.

Same tone. Same pacing. Same "I'm rushing" energy.

The message is short:

"Hey, I'm in the middle of something. We need to send the first tranche today or the deal stalls. I'm forwarding the wiring details. Please handle it now and don't escalate; this is sensitive."

Thirty seconds later, an email arrives with:

- The vendor name (a real one)

- The invoice number (real format)

- The right approval language

- "Urgent" but not messy

- A clean PDF that looks normal

No malware. No broken systems.

Just a request that fits the culture:

move fast, don't create friction, handle it.

The manager hesitates for half a second. Then does what good employees are trained to do:

They execute.

The money is gone in minutes, routed through accounts that don't look suspicious until it's too late.

And the part that hurts most isn't the money.

It's the feeling afterward:

"I did everything right. I wasn't careless. It just felt real."

What actually happened

This wasn't "hacking."

This was authority + urgency + realistic communication, delivered with AI precision.

What would have stopped it

Not "better instincts."

A process that can't be bypassed by a believable message.

- Two-person approval for wire transfers
- Mandatory call-back verification using a known number for bank detail changes
- A culture where "I'm verifying because it's my job" is respected, not punished
- A rule: no payment action based on email/chat alone

Lesson: the future of fraud is not louder, it's smoother.

War Story 2: The Hospital That Couldn't Open Doors

A hospital gets hit at 2:40 a.m.

At first it looks like an IT issue:

- Servers slow

- Printers stop

- Logins fail

Then it spreads into real life:

- The medication dispensing system won't authenticate

- The patient check-in process breaks

- Internal phones glitch

- The building access system starts failing in certain wings

- The backup network is "reachable" but useless

In the morning, nurses are doing what they can with pen and paper. Doctors are frustrated. Patients are scared. Leadership is trying to stay calm.

Then the extortion message arrives.

But it doesn't say "pay to decrypt."

It says:

"We already have patient data.

We already have staff information.

We already have internal emails.

We can publish proof today."

And the hospital leadership realizes the nightmare:

Even if they restore systems, they can't restore privacy.

Even if they recover operations, they can't recover trust.

Even if they refuse to pay, the threat isn't only downtime.

It's exposure.

It's lawsuits.

It's regulatory pressure.

It's headlines.

What actually happened

This wasn't just ransomware.

This was extortion-first, combined with disruption, hitting not just computers, but the systems that make the building function.

What would have limited it

Not one magic tool.

A set of boring, powerful decisions made before the crisis:

- Segmentation so building systems and clinical systems don't share the same blast radius

- Backups that are protected from tampering and tested for restoration speed
- A practiced incident plan: who decides, who communicates, what gets isolated first
- Strong identity protection (because many incidents start with access, not exploits)
- A data minimization mindset: don't store what you don't need forever

Lesson: the future attack isn't just "files encrypted." It's operations + reputation + leverage.

War Story 3: The Breach That Waited Five Years

A professional services firm has a breach in 2026.

They handle it reasonably well:

- they rotate credentials
- clean systems
- harden controls
- communicate carefully
- move on

They assume the worst is behind them.

Then, in 2031, a journalist calls.

"Can you comment on your client files being circulated?"

The firm's leadership is stunned:

"We haven't had a breach since 2026."

The journalist sends proof:

- Contracts
- Negotiations
- Archived emails
- Documents that were encrypted at the time

That's the twist.

The attacker stole encrypted archives years ago, stored them quietly, and waited until cracking them became feasible.

No new breach required.

No new entry point.

Just time.

What actually happened

This is "harvest now, decrypt later" in real life.

Steal encrypted data today. Break it tomorrow, when the math becomes less protective.

What would have reduced the damage

This is where "future cybersecurity" becomes quiet and serious:

- Identify long-life secrets (data that must stay private for years)

- Create a crypto inventory (where encryption exists, what depends on what)
- Design for crypto agility (systems that can upgrade without rebuilding everything)
- Don't keep sensitive archives forever "just in case."
- Pressure vendors and partners: what is your long-term encryption plan?

Lesson: Sometimes the breach doesn't end when you kick the attacker out. Sometimes the breach ends years later, when the data becomes readable.

War Story 4: The One-Second Storm

A mid-size company launches a new product.

The website is backed by a decent cloud provider. They feel safe.

At 10:03 a.m., traffic spikes.

By 10:04, the site is down.

By 10:05, customer support is overwhelmed.

At 10:06, leadership thinks it's just a DDoS.

At 10:12, the real attack begins: while everyone's distracted, attackers target access workflows, password resets, admin portals, and helpdesk tickets.

The DDoS isn't the main event.

It's cover.

By the end of the day:

- Customer records are accessed

- Internal tools are tampered with

- Social accounts are hijacked

- A ransom note arrives offering to "stop the storm."

The company learns the new reality:

Some attacks are not about "breaking in."

They're about exhausting you until you make a mistake.

What actually happened

This was a blended attack:

- Disruption + distraction

- Identity pressure + workflow abuse

- Speed as the weapon

What would have changed the outcome

- DDoS mitigation plans that exist before launch day

- Identity hardening and "no risky action from email alone" rules

- Rate limits, alerting, and approval workflows that don't collapse under pressure

- Runbooks: when the storm hits, everyone knows their role
- A culture that treats "pause + verify" as strength, not slowness

Lesson: the future isn't one threat at a time. It's stacked pressure designed to break your decision-making.

The Pattern Behind All Four Stories

Different targets. Different techniques.

Same core truth:

Cybersecurity is becoming a fight over trust, speed, and leverage.

And the way you win isn't by trying to predict every single attack perfectly.

You win by designing your life and your organization so:

- One believable message can't move money
- One stolen login can't reach everything
- One incident can't erase your ability to operate
- One crisis can't force panic decisions

A big reason these stories feel realistic is that many breaches still begin with social engineering and phishing, attacking the human layer instead of the hardest technical layer.

You've read four stories.

Different settings. Different victims. Different methods.

But they all had the same heartbeat.

Not "hacking."

Trust. Speed. Leverage.

That's the future of cybersecurity in plain language.

And if this book did its job, you don't feel helpless right now.

You feel clearer.

Because the goal was never to scare you with the future.

The goal was to **prepare you for the reality that's already arriving.**

The Pattern Behind Every War Story

Let's take one line from each scenario:

- **The Voice That Approved the Wire:** the attack wasn't technical; it was believable.

- **The Hospital That Couldn't Open Doors:** the attack wasn't just a disruption; it was pressure.

- **The Breach That Waited Five Years:** the attack didn't need to win today, it just needed time.

- **The One-Second Storm:** the attack wasn't one threat; it was stacked threats designed to exhaust you.

So, the future-proof response is the opposite:

Slow the risky moments down.

Spread trust across more than one person.

Build systems that don't collapse under pressure.

The "New Rules" of Cybersecurity

(If you remember nothing else, remember these)

1. **Urgency is a red flag, not a reason to rush.**

2. **A message is not authorization.** (Systems authorized. People verify.)

3. **One person should never be able to move money alone.**

4. **One login should never be able to reach everything.**

5. **Recovery is part of security.** (Backups and plans aren't optional.)

6. **Assume a compromise at some point, design for blast radius.**

7. **Verification is professionalism now.** Not rudeness. Not mistrust.

8. **Your email is your master key. Protect it like a vault.**

9. **Vendors don't deserve unlimited trust just because they're "big."**

10. **You don't need perfect prediction. You need strong defaults.**

A Final Note

If you've read this far, you already know something important:

The future of cybersecurity isn't about fear.

It's about awareness, preparation, and calm decisions when things feel urgent.

The threats will keep changing.

The tools will keep evolving.

The stories will keep getting more believable.

That part is out of control.

What is in our control is how we respond.

You don't need to be perfect.

You don't need to predict every attack.

You don't need to live in suspicion.

You just need habits that slow down risky moments.

Systems that don't depend on flawless judgment.

And the confidence to verify instead of rushing.

If this book helped you pause, question something that "felt normal,"

Or design your life or organization to be more resilient, then it did its job.

Stay curious. Stay calm. Verify when it matters.

That's how you stay safe in a world where trust can be faked.

www.ingramcontent.com/pod-product-compliance
Lightning Source LLC
Chambersburg PA
CBHW040857210326
41597CB00029B/4873